Robert Hebert Quick

How to Train the Memory

The Three A's

Robert Hebert Quick

How to Train the Memory
The Three A's

ISBN/EAN: 9783744724647

Printed in Europe, USA, Canada, Australia, Japan

Cover: Foto ©berggeist007 / pixelio.de

More available books at **www.hansebooks.com**

HOW TO TRAIN THE MEMORY.

THE THREE A'S.

By R. H. QUICK,

AUTHOR OF "EDUCATIONAL REFORMERS."

NEW YORK AND CHICAGO:

E. L. KELLOGG & CO.

1891.

HOW TO TRAIN THE MEMORY.

THE THREE A'S.

WITH the object of illustrating the connection between the theory and practice of education, that is, between the conception of what is to be done, and the means of doing it, I offer the following remarks on Memory, and its treatment in the school-room.

Whatever we see or become conscious of by way of our senses has an effect upon our minds; also everything that we think or wish. Whether that effect is in all cases indestructible is not a **The mind holds its impressions.** settled point, though some very singular occurrences have proved that we retain far more than we ourselves suppose. A remarkable case has been reported from one of the London hospitals, of a man who in the delirinm of fever suddenly began to speak in an unknown tongue. The language was at last identified as Welsh. When the man recovered, he said that he had spoken Welsh when a boy, but had since lost it, and could not when in health remember a word of it. So it may be that all impressions are permanent; but however this may be, our minds retain the *residua* of a vast number of impressions, many more than we can remember or recall at will. When a past impression returns to our consciousness, we are said to have an " idea," or a " re-presentation," of that impression. These "ideas" are seldom perfect. They may be very faint, and although they return to our consciousness when suggested by some similar impressions or ideas, we may have no power of recalling them by an effort of the will. And when they do come back to consciousness, they may be incomplete, or even partly incorrect. Suppose, e.g., I hear a name for the first time: to-morrow I might be unable to recall it, though

Examples. a similar sound might suggest it to me without my wishing it. If I wished to recall it, I might produce a sound somewhat like it, and not feel quite sure whether this was exactly the name or not. I have taken here a very simple instance. In other cases the " idea" *must* be incomplete, if not incorrect. When we have seen a picture that has interested us, we retain an impression that will for a time give us an "idea" of the picture, though an imperfect one.

The power of recollection, or bringing readily into consciousness correct ideas of past impressions, is a power which may be indefinitely increased by judicious practice. Teachers know this well, and to this power of the mind, at least, they attach due importance. In every school-room, then, much time and energy are devoted to this "cultivation of the memory." But we should probably succeed better if we attended a little more to theory, and studied the nature of the faculty we wished to cultivate.

There seem to be different kinds of memory, so to speak. One person can remember words, another numbers, another places, another never forgets a face. These different kinds of memory depend partly on natural ability, partly on training. At Trinity College, Cambridge, there are in residence over 400 undergraduates, of whom rather more than 100 change every year. Yet the porters who have to know every one by sight, very rarely ask a name more than once. Still more extraordinary is the way in which they remember every one's address. They are in fact perfect walking address-books. Knowl-edge of this kind is mostly kept as long as it is wanted and then thrown over; but when I re-turned to Cambridge after an absence of ten years, I was amused to find that the porter re-membered the letter of the staircase on which I had lived, although till reminded by him, I doubt if I could have told this myself. Illit-erate people are sometimes so thrown upon the resources of their memory that, from exercise, this becomes extraordinarily powerful. There have been cases of such people doing a great deal of complicated buying and selling, and trusting to their memories with as much success as other people trust to their account books. The way in which the memory is strengthened by habit is, I think, well illustrated by the following anecdote, trivial as it is in itself. A country postman once told me he was in the habit of getting an occasional lift in a butcher's cart, and he then saved the butcher's man trouble by taking orders for him at houses that lay off the road. When several things were ordered, he had some difficuly in keeping them exactly right in his head till he rejoined his friend; but the butcher took not only these orders, but orders at houses for miles round, and without difficulty kept them all in his head till he went back to the shop; nor did he ever make a mistake, however numerous the orders might be.

I have given these instances to show you how memory is developed by practice. And you will observe that the general memory is not strengthened by these special developments. The college porter and the butcher are like other men, except with reference to the special class of facts they have to remember.

Neglect of these very obvious truths has led

to much injurious action in the school-room. The maxim of the old scholars was that so often repeated by Casaubon—" *Tantum quisque scit quantum memoriátenet :* Every man knows just what he remembers." The modern school-master in this, as in other matters, has taken his cue from the old scholars. But for want of careful discrimination of the different kinds of memory he has often developed a kind of memory which is least valuable, if indeed it is not injurious to the other powers.

We must all have met with the following experience. We are engaged in thought, when a question on some subject not **Sensational** connected with our thoughts is **memory.** put to us. For some seconds we go on thinking, and though retaining the sound of the question, we are quite unconscious of its meaning. We then turn our attention to it and, as it were, read off the meaning from the idea of the words which we have retained. From this, we discover that the sensational and intellectual functions of the brain are pefectly distinct. Each of these functions has its peculiar kind` of memory, and it would seem that sensational memory may be developed at the expense of the intellectual. Certainly the two do not necessarily grow together, and stupid people and even idiots often have great power of sensational memory, *i.e.*, memory for mere sounds;

with which we may classify the memory of facts
retained without consciousness of their connection with other facts or with one another.   This
probably has given rise to the French proverb—
Beaucoup de mémoire et peu de jugement, and Pope
says :

> " Thus in the soul while memory prevails
> The solid power of understanding fails. "

I have known some singular instances of the
strength of this kind of sensational memory in
persons of weak intellect.   I have met with the
case of a lad who, though he knew nothing else,
knew the times of arrival and departure of most
of the trains in and out of London, which he
studied every month in Bradshaw.   A pupil in
a school where I was master had a remarkable
faculty for learning by heart, though he was
very dull in other respects; and his memory was
so purely sensational, that when he was set to
learn the Kings and Queens of England with
dates, so as to be able to write them down, he
learnt the list of Kings and the list of dates sep-
arately, and wrote them without endeavoring to
connect them in his mind; *i.e.,* he wrote first the
list of Kings without thinking of dates, and then
the list of dates without thinking of Kings.   We
discovered this by one of the dates having dropt
out, so that in modern times the kings did not
come to the throne till their death.   On being

asked about this, he explained his mode of procedure.

Now, it cannot be denied that most school teaching of children tends to cultivate the sensational memory mainly, if not exclusively. The school-master wants some ostensible *examinable* results of his teaching, and he gets this most easily by making his pupils simply learn by heart. There is a tendency in both teacher and pupils to make learning go easily, so to say, and exercise soon gives great power to the sensational memory; so that if it is not over-driven it jogs along with much satisfaction to its possessor and his teacher. As Brudenell Carter has well said, the child who uses his sensorium to learn words, is using an instrument perfected for him by the great Artificer; but when he comes to use his intelligence, he no longer uses a perfect instrument, but a faculty which is as yet only partially developed. He cannot therefore use it so easily. He must make an effort and puzzle his head before his intelligence will act at all. I was lately hearing some children say tables. " What is 7 eights ?" I asked, and got the prompt answer, " 56." " How many eights added together make 56 ?" I asked next, and no answer at all was forthcoming. The first question was addressed to the sensational memory, the second to the intellectual. Another

Most teaching of children cultivates the sensational memory.

(9)

instance occurs to me. A lady who had just
given a lesson in an elementary school to some
young children, told me she began to talk about
geographical definitions. "You know," said she,
"that an isthmus is a narrow piece of land joining two continents." "*Connecting*, teacher!"
shouted the children. "Very well," said she,
"connecting two continents. Now, who can tell
me what is meant by connecting?" and she
found that not a child had the smallest notion.

Now as things run far more smoothly when
the sensational memory only is exercised, we
Neglect of the cannot be surprised that so much
intellectual use is made of it; but the develop-
memory stupe- ment of this kind of memory and
fying. the neglect of intellectual memory
leads to the stupefying of our children. "They
won't think," complains the schoolmaster quite
pathetically. Why will they not? They think
about their games, about their schoolfellows,
about their masters, about their homes. They
think shrew.ly enough on these subjects, and
perceive many an error in the master, of which
he too might think with advantage; but about
school work, they certainly seem to have no power
or will to reflect on anything. Very much of this
comes from the common notion that the first
school lessons must exercise the sensational
memory. Children learn Kings and Queens,
capital and county towns, tables, parts of speech,

declensions, conjugations, and the like, and they are not expected to have any conception whatever to connect with these sounds; so they naturally acquire the habit of using in the school-room the sensational memory only, and when the habit is well established the luckless schoolmaster is appalled by their seeming stupidity.*

School work then, as a rule, makes too much of sensational memory. Next, it develops the carrying rather than the storing **School work** memory. The mind by practice **should develop** can acquire the art of rapidly **storing power.** getting up a lesson, and as rapidly forgetting all about it. This "carrying power" is especially useful to barristers and actors, and they perform

---

* I have been asked, " Do you then condemn learning by heart?" To which I reply: " No, but learning by heart is not all of the same kind." What I object to, is learning that exercises nothing but the sensorium. If the children are *interested* in what they learn, the sensorium is in no danger of being over-developed. But the general notion is, let words be learnt by heart first, and then the intelligence will play its part afterwards. I have heard of a schoolmaster who, in teaching his boys to read, enjoined them never to think of the meaning—that would only distract their attention. " One thing at a time is my maxim," said he. It is against this " unhappy divorce of words and things" (as Comenius calls it) that I wish to protest. If I cannot get a hearing as "theorist," I would appeal to results. The great difficulty of all schoolmasters is, that children, after the ordinary preparatory course, never look for a *meaning* in the words of the book. Surely " God's great gift of speech" must have been " abused," when learners no longer expect words to mean anything.

(11)

great feats of this kind.   Actors *study* parts they are likely to act often, but they *get up* a part that is wanted only for a special occasion, and a part thus got up is forgotten immediately after the performance.   The memory adapts itself wonderfully to circumstances.   A friend of mine, who has to review a great many books, tells me that when he has read a book he remembers all about it till the review is written, and then he gets rid of the subject from his mind as easily, and, as far as he knows, as completely as he gets rid of the book from his table. · Now the getting up of lessons fosters this habit of mind.   The mind has to lade itself with certain knowledge and "carry" it for a few hours, and then it drops it, not without a feeling of relief.   "The tear forgot as soon as shed" is a well-known characteristic of childhood, and so too is *the task forgot as soon as said*.   Unfortunately, our competitive examinations place a very high premium on the cultivation of this kind of memory.   I remember in a large school a prize was offered for the best examination in a certain set of books on a period of English history.   When the appointed day arrived some cause of delay arose, and it was announced that the paper would not be set for a fortnight.   One of the boys, who was very successful in such examinations, thought himself much injured by this alteration.   He had prepared himself, he said, for the day fixed, and in consequence of the change he would have to go all over the subject again; if he did not, in a fortnight's time it would have entirely gone out of his head. This carrying power is no doubt useful in some circumstances, but it is not memory, if we consider memory as the hoarding power of the mind; and its extreme development in the school-room is no doubt injurious.

We learn, then, that the schoolmaster, in trying to cultive the memory, too often cultivates the wrong kind of memory; first, that which is merely sensational, and secondly, that which is merely the carrying as opposed to the storing power of the mind.

The three A's.

How then should memory be cultivated? We should attend to what have been called "the three A's." These are ATTENTION, AR-RANGEMENT, ASSOCIATION.

1. The art of memory is the art of *attention*, said Dr. Johnson; and another thinker has declared that genius itself is nothing but the power of continuous attention. The mind's power of retaining an idea varies as each of the following three things —1st, the strength of the first impression, which strength depends on the whole mind's being concentrated on forming the, idea, in other words, on the amount of attention given it; 2nd, the length of time during which the thought keeps possession of the mind; 3rd, the

Attention.

(13)

frequency of its renewal, *i.e.*, the number of times it is brought back into consciousness. The first thing to be secured then is *attention*.

As we all know, there is such a thing as voluntary attention, when the mind resolves to fix itself on a certain subject and does so. We are constantly expecting young people to give voluntary attention to the work before them, and we say that the power of voluntary attention is of the very greatest importance. No doubt it is. But voluntary attention is one of the highest functions of the trained intellect, and nothing is more ridiculous than to make great demands on the voluntary attention of young people. It is, in fact, to expect at the outset of their intellectual training just what that training will *in the end* give them, where it is perfectly successful. In the early stages, we must think more of involuntary than of voluntary attention, and by means of it must cultivate a *habit of attending*. Even involuntary attention is not continuous in the very young. We see the infant attracted by some object, say a bunch of keys. In a few seconds it throws it away and grasps at a watch-chain. In a few seconds more it turns from this to look about for something else. Here we have the power of attention in the earliest stage of all; and in the next, *i.e.*, in young children, there is, as we all know, a restlessness which can be satisfied only by perpetual change in the direction of thought. If the teachers neglect this simple truth about the nature of the mind, unpleasant consequences are likely to ensue. The children will soon cease to attend even to instructions which for a little while may be well suited to them. When they are no longer occupied with the matter in hand they speedily become "naughty," that is, each child's energy takes an independent direction, and the harmony of the class is at an end. To restore it, the teacher has recourse to punishments, and thus from their earliest years children are accustomed to look upon learning as one of the chief troubles of life.

Instruction in its first stages then, should aim at securing the involuntary attention of the children, and should gently foster the increasing power and habit of attending to one thing without wandering. Later on, when the mind has some power of dwelling on a subject, pains must be taken to cultivate *voluntary* attention. There are studies especially valuable in this way, as *e.g.*, geometry; but the main thing is to get the whole mind concentrated on the work in hand, whatever it may be. This habit of concentration is fostered by letting school exercises and preparation be done without fixing a definite duration for the work. If boys have no inducement to get the work done soon, they will acquire a pottering habit, and their minds will wander; but if they may turn to occupations more pleasurable to them as soon as the work is completed, they will put out all their strength to come to the end. Over-hurrying is indeed likely to take the place of pottering, but it is perhaps the lesser evil of the two, or at least, the easier of correction.

But I have been considering continued attention generally, rather than intensity of attention at the outset, which is the cause of strong first impressions. Now intensity of attention, with the young at all events, depends entirely on that almost unaccountable thing which we call "interest." When the mind is interested, all its powers are ready for action; when uninterested, it seems in a state of coma. Whenever then we can arouse interest we are likely to impress the memory. The converse of this is recognized in the affairs of every day. Suppose, *e.g.*, an acquintance invites us to dinner and we, having accepted the invitation, forget the engagement and do not go; the reason of our non-appearance is regarded as an insult, and that for an obvious reason. Our forgetfulness is a proof that we were not much interested by the invitation, for if we had been we should not have forgotten it.

See page 36 for context.

* In an article on " The Memory and its Doctors," in the New York *School Journal*, Aug. 25, 1888, I see an account of Mr. Loisette's " Interrogative Analysis." This seems to consist in a series of questions on the piece to be learnt. By way of answer the whole sentence is to be repeated in the words of the original, and the answer conveyed by emphasizing the right word. A few good ques-

Similarly, in the school-room, if the master were to announce to the school " The French elections have been fixed for the 18th of October; try to remember that "—the chances are that the 18th of October would not remind a single boy of the elections. But if he said "On the 18th of October there will be a total eclipse of the sun, and it will be dark in the middle of the day,"— nobody would fail to expect this when the day arrived. And so we find everywhere that our knowledge, *i.e.*, the area brought within our ken by memory, spreads just where we take an interest and nowhere else. The first step then towards bringing about healthy exercise of the memory, must be the awakening of interest in the thing to be remembered.

But when a vivid first impression is once secured, the mind must dwell upon the idea before it is allowed to pass out of consciousness; otherwise speedy recollection will be impossible. We see this from the way in which

The concept must be dwelt upon.

novels are forgotten now that the supply is unlimited, and boys devour them in great numbers. Years ago, when novels were not easily obtained, we did not hurry over the feast, and our impressions were more lasting than those of the young novel-readers of now-a-days, who remind

---

tions might be formed and answered in this way, but if the plan were carried out after the Loisettean model it would speedily become a bore, and the rule generally holds—" Every way is good but the tiresome."

one of the old joke about reading *Ten Thou-sand a Year*.   In school-teaching, the concepts, when accurately obtained, are often not prop-erly dwelt upon, and it is no unusual thing for a master to finish off all the definitions with his first Euclid lesson.   He assumes that when once the concept is formed it will remain in the boy's head forever;   whereas it must be dwelt upon till the mind is familiar with it: and further, it must be brought back again and again into con-sciousness, so that it may present itself uncalled-for whenever it is wanted.   For in the mind well furnished and well trained, the ideas will deserve the eulogy pronounced by James I on his courtier Sir Henry Wotton: They will never be *in* the way, and they will never be *out* of the way.

This brings us to the third thing necessary, viz., frequent repetition.   All great authorities in school matters are agreed on the necessity of a good foundation, *i.e.*, of knowing thoroughly the things taught first.   There is indeed, a great difference in the various notions about knowl-edge.   Some people mean the exercise of the sensational memory only; others, like Pesta-lozzi, mean thorough grasp of elementary ideas. Some teachers, again, require in every subject thorough mastery of tables by the sensational memory, and at the same time full play of the intellectual memory about ideas which the tables serve to suggest and connect. But all alike require that the ground should be gone over again and again till the recollection, and bringing the idea back into consciousness, takes place without effort. Only then has the knowledge become a part of the mind's available property. The following amusing passage from an admirable little book, Jacob Abbott's " Teacher," puts before us very clearly the difference between the perfect and the partial action of the memory:—

"Can you say the Multiplication Table?" said a teacher to a boy near him in class. "Yes, sir," said he promptly. "Begin at 9 × 1" said the teacher. The **Example.** boy went through the 9's slowly but quite correctly. "Begin again," said the teacher, "and I will try an experiment. Mind you don't stop till you get to the end." Directly the boy had begun the 9's the teacher also began saying aloud the 7's. The boy went on a little way and broke down. "I know the table, sir," said he, "but I can't say it because you put me out." "Very well," said the teacher; "say the Alphabet." Directly he began, the teacher started also, beginning at another place, but this time the boy went on to the end without difficulty. "You see, now," said the teacher, "that though you know both the Multiplication Table and the Alphabet you know them in very different ways."

Now the things which the mind will have to use frequently we want thoroughly mastered, and this cannot be secured without frequent repetition. But then arises one of the teacher's greatest difficulties. The mind, especially the mind of the young, will enter into nothing in which it is not interested; and mere repetition is a deadly foe to interest. How then is interest to be kept up while ideas are brought back into consciousness often enough for the mind to be able to recall them without effort ? The true secret is, as I believe, to make as little use as possible of merely sensational memory, and to vary the mode of bringing the idea back to the mind. Take, for instance, the Multiplication Table, which is learnt and perhaps must be learnt at first by the sensational memory: it is easy to ask questions in a variety of ways so as to set the mind to work upon it. Suppose, *e.g.*, the 4 line is known, the teacher may ask, If I take 4 and 4 and 4 and add them together, how many 4's should I have ?—what will that make? If ten 4's are 40, and I take away 4, how many 4's are left ?—how many would that be ? When the children are more advanced they may say tables in a variety of ways, *e.g.*, the teacher may say, Name all the multiples of seven less than 100. Name the odd multiples of nine under 100. Go up all the numbers to 100 and say which are prime numbers and which are multiples. Exercises of this sort teach pupils not only to recollect with ease, but also to use the truths recollected.*

In his efforts to get variety in the manner of repetition without changing the substance, the teacher should employ the various senses wherever this is possible. **Need of variety in repetition.** The ear, the voice, the eye, the hand, may often be exercised about the same matter. The effect of using more senses than one is in itself a capital thing for the memory. The idea formed by the action of the two senses is stronger than that formed by the action of one. To test this, you may try the experience of seeing how much of a printed sentence you can take off by reading it to yourself and then writing it without book, and how much you similarly take off when you read the passage aloud. You will find that the eye and ear together are stronger than the eye alone.

2. We next come to the ASSOCIATION OF IDEAS, which as James Mill long ago pointed out, is a powerful instrument in the hands of the thoughtful edu- **Association.** cator; for by this association of ideas, one idea,

---

* We must not forget, however, that brain-work takes time; and no one without experience in teaching would believe how often the mind has to connect 9 × 6 with 54 say, before the first immediately suggests the last. The necessary amount of repetition could perhaps hardly be secured if we *always* associated brain-work with it,

as a matter of course, suggests another, and the mind tends to form established trains or sequences. These sequences are under the influence of custom, of pleasure, and of pain; and all these depend in some measure on the educator. As we are now considering the memory, only, I will not discuss the larger question of habit, which is a result of the tendency in both mind and body to act in established sequences; but in passing, I cannot help remarking on the folly of associating in the minds of children pain and disgust with the things which we wish them to become attached to. As Locke says, the very sight of the cup from which we have been accustomed to take nauseous physic is unpleasant to us, and we can relish nothing we drink out of it. Why, then, do we so often make books instruments of torture to children, especially the children of the poor, if we do not wish them to hate the sight of a book all their lives ? Why do those who love religion so often inflict tedious religious services on children, unless they wish the children to shun religious services as soon as they are their own masters ?

But this by the way. We are considering association of ideas as a help to memory. The singular ease with which the mind runs along established trains may be readily tested by saying the Alphabet forwards and then trying it backwards. I do not know, by the way, why this particular train is so well established in all of us, unless it be that it was one of the first sequences of any length to which the mind became accustomed.

Now our knowledge, in order to be of any use to us, must not lie in the memory, a pile of isolated facts, but must be worked up into trains along which the mind will work without effort. In the words of an old writer; " There are persons who have laid in vast heaps of knowledge which lie confusedly and are of no service to them for want of proper clues to guide into every spot and corner of their imagination; but when a man has worked up his ideas into trains, and taught them by custom to communicate easily with one another, then arises order, and he may reap all the benefit they are capable of conveying; for he may travel over any series of them without losing his way and may find anything he wants without difficulty." (Abraham Tucker's *Light of Nature.*)

We see now how the teacher may strengthen the pupil's memory. He must not require them, as the authors of most school-books do, to perform the *tour de force* of committing to memory **Practical suggestions.** a huge number of disconnected facts, but he must awaken in them a perception of all *the connecting links* between what is already known and what is to be remembered. Mnemonics, as you know, give purely arbitrary connections between the things to be remembered. This sort of connection is better than none at all, but it is far inferior to connections which lie in the things themselves. When anything new is to be received, the pupils should be led to compare it with what *they already know* and to mark similarities and differences. Too often, pupils are raced along and made to acquire imperfectly, by sensational memory only, a large quantity of sounds; and similarities which might be a great assistance to them become a mere source of confusion. *E.g.*, a boy learns the verbs in the the verbs in the Latin grammar from the beginning of the active of *amo* to the end of the passive of *audio*. In this case, things which should be for his wealth prove an occasion of falling; for the similarity between the conjugations, and between active and passive voice, leads to all kinds of wrong combinations. But if the active of *amo* is made familiar to the learner and he has then to learn the passive of *amo* or to go on to the active of the next conjugation, he may compare what he knows with what he has to learn, and by this means may materially lighten his labor. School-masters in large schools have a similar experience in remembering boys. If two boys a good deal alike enter the school at the same time, the masters often go on confusing the one with the other; but if a boy enters the school, who is a good deal like another whose face has already become familiar, there is no confusion, because the masters think of him as the new boy who is so like the boy they already know.

Before I quit the subject of connection of ideas, I must give a caution which we all stand in need of. By the time we have grown up, we have formed in our minds all kinds of trains of ideas, and by habit we have got to think of these associated ideas as if they were one simple idea; and hence we attribute to other people, often indeed to our pupils, the possession of the whole connected series, when they have but a part. We expect them, too, to keep up with us when we are going along a well-worn high road, and they are, so to speak, on the other side of the hedge and have to scramble along over a very rough country. A little more knowledge of the operations of the mind would cure a good deal of the shool-master's impatience.

3. The last of the three A's, ARRANGE-MENT, is closely related to the second, Associa-tion. When things are well ar-ranged, the mind can form good **Arrangement.** trains of ideas; and natural connections, as I have said, are far better than artificial; indeed, memory of real connections is the memory of great in-tellects, memory of isolated facts is the memory of idiots. Very great care then should be taken

by the teacher to put the different things to be retained in good order.   In Thomas Tate's "Philosophy of Education" is the following story, which well illustrates the power of arrangement in assisting the memory:—*

"Betty," said a farmer's wife to her servant, "you must go to town for some things. You have such a bad memory that you **An Example.** always forget something, but see if you can remember them all, this time." "I'm very sorry, ma'am," says Betty, "that I have such a bad memory; but it's not my fault; I wish I had a better one."   "Now mind," said her mistress, "listen carefully to what I tell you.   I want suet and currants for the pudding." "Yes, ma'am, suet and currants."   "Then I want leeks and barley for the broth; don't forget them." "No, ma'am, leeks and barley; I shan't forget." "Then I want a shoulder of mutton, a pound of tea, a pound of coffee, and six pounds of sugar. And as you go by the dressmaker's tell her she must bring out calico for the lining, some black thread, and a piece of narrow tape." "Yes, ma'am," says Betty, preparing to depart.   "Oh, at the grocer's, get a jar of black currant jam," adds the mistress.   The farmer, who has been quietly listening to this conversation, calls Betty back when she has started, and asks her what she is going to do in the town.   "Well, sir, I'm

---

* I have not quoted with verbal accuracy.

(26)

going to get tea, sugar, a shoulder of mutton, coffee, coffee—let me see, there's something else." "That won't do," said the farmer; "you must arrange the things, as the parson does his sermon, under different heads, or you won't remember them. Now you have three things to think of—breakfast, dinner, and the dressmaker." "Yes, sir." "What are you going to get for breakfast?" "Tea and coffee and sugar and jam," says Betty. "Where do you get these things?" "At the grocer's." "Very well. Now what will be the things put on table at dinner?" "There'll be broth, meat, and pudding." "Now what have you to get for each of these?" "For the broth I have to get leeks and barley, for the meat I have to get a shoulder of mutton, and for the pudding I must get suet and currants." "Very good. Where will you get these things?" "I must get the leeks at the gardener's, the mutton and suet at the butcher's, and the barley and currants at the grocer's." "But you had something else to get at the grocer's?" "Yes, sir, the things for breakfast—tea, coffee, sugar, and jam." "Very well. Then at the grocer's you have four things to get for breakfast and two for dinner. When you go to the grocer's, think of one part of his counter as your breakfast table and another part as your dinner table, and go over the things wanted for breakfast and the things wanted for dinner. Then you will

remember the four things for breakfast and the
two for dinner. Then you will have two other
places to go to for the dinner. What are they?"
" The gardener's for leeks, and the butcher's for
meat and suet." "Very well. That is three of
the four places. What is the fourth?" "The
dressmaker, to tell her to bring out calico,
thread, and tape for the dress." "Now," said
her master, "I think you can tell me everything
you are going for." "Yes," said Betty; "I'm
going to the grocer's, the butcher's, and the
gardener's. At the grocer's I'm going to get
tea, coffee, sugar, and jam for breakfast, and
barley and currants for dinner. But then I
shall not have all the things for dinner, so I
must go to the butcher's for a shoulder of mutton and suet, and for leeks, to the gardener's.
Then I must call at the dressmaker's to tell her
to bring lining, tape, and thread for the dress."
Off goes Betty and does everything she has to
do. "Never tell us again," said her master,
"that you can't help having a bad memory."

I hope I have by this time shown you that even
such imperfect science as we have ought to influence practice in the schoolroom.
We have seen that there are dif-
ferent kinds of memory. The sensational action
of the brain has its memory, and the intellectual
has its memory. We have our choice, to some
extent at least, which kind of memory we will

Summary.

develop in our pupils, and we mostly develop that which works easiest, the sensational. Science would teach us that this is wrong, and that we should endeavour to make intellectual memory take the place of sensational. Next we found that the mind has two very distinct powers, which may be called the carrying and the hoarding power. The carrying power has its uses in special circumstances, and can never be neglected so long as there are examinations to prepare for; but the hoarding power is one of the principal faculties of the mind, for the intellect without a hoarded treasure of truth works to little purpose, as a flour-mill with no corn in it. The mind then must be taught not how to carry, but how to hoard; and for this purpose we must cultivate its interests, we must accustom it to continued attention, we must teach it how to arrange its ideas and connect them in trains, so that one idea may call up others bearing on the same subject.

Perhaps the gist of what I have said will be seen most clearly if we take a subject and see how the previous considerations will affect the teaching.

Learning poetry has always occupied a large place in the curriculum, though till quite lately the poetry learnt in our great **Practical suggestions.** Has any attempt been made to secure the right

sort of memory in this case? Very seldom, I
believe. We always go back to our own childhood and make our own experience the test of
the general experience; and adopting this plan,
I call to mind the time when on joining a new
class I began in the middle of Gray's Ode and
learnt:

> " Alas, regardless of their doom,
>   The little victims play;
> No thought have they of ills to come,
>   No care beyond to-day."

I very well remember puzzling myself by trying
to think who the little victims could possibly be,
what their doom was, and why they didn't mind
it.   Still in this case I hoarded the words, and
some eight or ten years afterwards I managed
to attach some meaning to them; but on being
moved to a public school (Harrow) I found that
the carrying power was the truly valuable one.
I wanted to " get my remove," and I found my
getting it would depend in a great measure on
the quantity of Ovid I could say by heart.   I
therefore managed to carry in my head for a
little while a great quantity of verses, of which
I never attempted to construe a dozen. I got
them up by parrot memory only: they were
nothing but sounds, and oddly enough it was
an understood thing that we were not required
to know the meaning ! In this case great importance was attached to memory, but to
memory of the wrong kind. This was at Har-

row. At another public school (Winchester), in days gone by, there was an attempt made to cultivate both the hoarding and the intellectual memory by the following expedient:—In every examination while he remained in the school a boy might take up the Latin and Greek repetition which he had prepared for his first and subsequent examinations, so that he gained a store which was kept and increased as he went up the school. Thus the hoarding memory was encouraged. Besides this, he was not allowed to say anything he could not construe, so here was some precaution taken against mere sensational memory. In a little book published some years ago by the Rev. Henry Fearon, he says that he knew at Winchester a boy who could construe and repeat 14,000 lines from Latin and Greek poets.

This Winchester plan had some very good points in it; but as the boys were left to learn up the repetition in their own way, the great probability is that in learning by heart they had little consciousness of the meaning, for both young and old have a tendency to avoid thinking; and in a foreign language the sounds do not so readily suggest ideas as in our own language. I remember asking a lad if he ever thought of the meaning when he repeated Latin poetry, "Yes," he said, "sometimes—*when I can't think of the Latin.*"

For this and other reasons good pieces of English poetry should be *learnt*, that, is not car-

As to Poetry. ried for a few days but hoarded for life. For this purpose they must be much more elaborately studied than poetry usually is. The ordinary way is for the teacher to set so much to be got up, and the children then read it over and over till they can "say" it. Sensational or parrot memory is therefore used at first if not at last also. True, many teachers will say; but this must be the case here as in almost all learning. Your Inno-vators would have nothing learnt by heart with-out full understanding; but full understanding is seldom possible. Who can say that he fully understands the highest utterances of great poets and thinkers? Are we then to learn only the inferior things which we can perfectly un-derstand? And if you admit that the child can understand very little perfectly, you must admit that he should learn what he does not under-stand: in other words, you grant him the use of his sensational memory.

In reply to this, I contend that it is the educa-tor's business to develop the memory which is most important and least able to take care of itself. It is indeed true that comprehension, even in adults, is far from perfect, and in children it is very imperfect indeed; but instead of as-suming that children can't understand, and so getting them accustomed not to expect sense, the educator should train them to endeavor to understand. The child, when he begins to learn, will be ready to say with the Student in *Faust :* "Ein Begriff muss bei dem Worte sein—The words must surely have a meaning." · But the schoolmaster too often answers like Mephistopheles : " Schon gut! Nur muss man sich nicht allzu aengstlich quaelen—No doubt they have, but you need not bother yourself about it." The educator will try to make children discontented till the words have a meaning for them.*

Remembering that the mind works only where it is interested, the master will choose a piece of established excellence, simple in its character, and of such a nature that it may connect itself with **Illustration of learning a poem.** what the children already care about. The children must *like* the piece. And it is interesting to the teacher to find what they like best. I have often tried the following plan with great success. I have selected six or eight poems which I knew were thoroughly *good* and suitable for the children. Everybody then has ` a paper and pencil. The teacher then reads a

---

* It is a most interesting question how far children who have not suffered from "teaching" do actually expect words to have meanings. At first they learn only the words they want, and every sound they acquire has its meaning : but they soon get to like jingles as such. I am by no means sure that the child is always so *exigeant* as Goethe's student.

piece to the class, and everyone (the teacher included) awards marks to it, 10 being the highest
possible. When as many pieces have been thus
read and marked as time will allow, the class
read out in turn the marks assigned, the teacher
giving *his* marks last. He thus finds which
pieces are the most popular, and the children
are much interested in comparing their estimates
with his.

He selects some piece which he finds popular,
say Cowper's poem " The Loss of the *Royal
George*," which is sure to be a favorite. As I
have said, a careless master will simply set the
piece to be learnt: a careful master may make
the opposite mistake of preparing a great quantity of information and trying to enforce it on
his pupils' memories—the date of Cowper's birth
and death, his melancholy, his friendship with
the Unwins, and much else which is not at all
to the purpose. All this literary information
does not interest the young and is never ac-
quired by them except for the examiner. But
the master may ask the boys about ships, about
the difference between merchantmen and men-
of-war, about the size of men-of-war and the
number of their crews, about Portsmouth and
its advantages as a harbour. I say he will *ask*,
for it is better to get information from the boys,
or at least their conception, which will have
been formed on all subjects that interest them;

and it is a good rule that the master should always talk as little as possible. The master may then tell the story of the disaster. He will say that this event was not in itself of such great importance as some other similar misfortunes, as *e.g.*, the loss of the *Captain*, but it has become celebrated through a poem. He will then recite the poem to them. He will next take a verse at a time and ask questions about the meanings of the words and phrases. He will ask especially for any incident of the story which is referred to in the poem ; *e.g.*, after reading the verse beginning "A land-breeze shook the shrouds," he will ask, On what coast of England is Spithead? What wind was it that upset the *Royal George?* And afterwards, with reference to the line, "His fingers held the pen," he will ask, How was the Admiral engaged when the accident happened?

A remark suggests itself to me about questioning. I think it will be well worth the master's while to have *thought out* most of his questions beforehand, and to have marked his book in such a way that a glance will tell him what questions he purposed asking. Next, if he asks the class collectively, two or three boys will answer, and the rest will feel they have no chance and will think of something else. If, on the other hand, he passes questions, a good deal of time is

As to questions
on the poem.

wasted; and besides, the first boy asked has not so much time to think as the last boy to whom the question descends: moreover, the last boy asked may have got some hint from previous guesses. Perhaps the best way is this: after asking a question and pausing the time requisite for thought, whether one second or twenty, to glance down one's list of the boys' names and stop the pencil at some name which one pronounces; if its owner is not ready with the right answer, the master answers for him and gives him a negative mark; but if he answers right, the master gives him a positive mark; if the answer is partly right, a mark may be given equivalent to O. In this way, the attention of the whole class is kept up. The marks cannot be made to give a fair result at the end of each lesson, and they should not be added together till after a series of lessons, when many questions have been asked.

Before the class have the poem to learn, they should have heard the master recite it on more than one occasion, and they should also have read it aloud to him. At this stage, attention may be called to the epithets by such questioning as this: "What is the shore which they were near called?" "Their *native* shore." "Why called native?" "The poet says she had sprung a leak. What kind of leak does he mention?" "A *fatal* leak." "What does this mean?" *

---

* See page 16 for note.

(36)

The main difficulty in learning poetry is to remember the order in which the verses come. The master should be careful to make the pupil observe any connection of thought in the consecutive verses. If the poem is a good one, the *fitness* of the order will come out on examination, and the perception of this fitness will assist the intellectual memory. The principle of association of ideas may be turned to account in another way also. Instead of reading one verse over and over, read always *two* verses. Read together several times the first and second, then the second and third, then the third and fourth. This way of forming a chain has been developed by Dr. Pick, and made the basis of many ingenious experiments.

In hearing the piece, the master should not prompt by giving the next word, but he should give the *sense* of what follows, and in this way lead the boy to depend on his thinking-memory.

When the piece is known, it must be recited very slowly and distinctly and with strict attention to the meaning. The boy reciting should stand as far as possible from the master. It very much enlivens these recitations (which take too much time to occur often) if the boys all mark the reciter and read out the marks, the master announcing *his* last. The boys will take great pains in their endeavor to get their marks near the master's.

We will suppose this and other pieces to have been learnt. In many schools, pieces of English poetry when once learnt are never thought of again. In these schools, the only things which are learnt to be remembered are Latin and Greek grammars. But good English poetry is at least as well worth remembering as doggerel verses about Latin genders. Let it be understood then, that the poetry will be useful again and again in school work. From time to time pieces may be written from memory. Sometimes the most emphatic word in each line may be underlined in these written pieces; sometimes the subject in each clause; sometimes the epithets; sometimes the prepositions or adverbs; and so on. Or the pupils may be required not to write the whole piece, but to write in column a list of the prepositions in it, with the words governed by them. Or the pupils may be told to mention any similes that occur in such and such a piece which they have learnt. Then papers may be set which will test not only the verbal, but also the intellectual knowledge of the poems. *E.g.*, "State everything that you can make out from the poem itself about the burial of Sir John Moore." Sometimes a question that can be more briefly answered will test intelligence as successfully. Take for instance, Charlotte Smith's *First Swallow*. In the first verse she writes—

" The oaks are budding, and beneath,
The hawthorn soon will bear the wreath,
The silver wreath of May." .

I lately asked, "In what month was the poem written? Give reasons for your answer." Almost all the boys answered, "May, because the wreath of May is mentioned." But the more thoughtful said, "April, because the swallow had just come, and the hawthorn would *soon* have the wreath of May."

Questions about the meaning and connection of different sentences are most important, because if the boys understand the words in connection, they cannot be altogether wrong about the meaning of the separate words. Besides, it is a great matter to make them attend to the thought expressed by the whole sentence. Everyone who has taught knows the tendency to disintegrate sentences, and give a meaning to words or clauses which the least thought of the context would prove to be untenable; as *e.g.*, in the fearful case, lately mentioned by an inspector, of a boy's explaining " his native air" as " the 'air of his own 'ead." But it would be very good for all of us, young and old alike, if we had to give an account of the exact sense in which we use words. I have heard it said of a songstress that she had a nice voice, but her singing did not give pleasure, because she was " seldom in the middle of the note." I am much mistaken if scrutiny would not show that our words are often like the sounds produced by this lady, and that we are not in the middle of the meaning of them. The young are specially likely to form false associations of words and meanings; as in the case of the boy who was asked the meaning of *wholesale* and replied that it meant *retail*. I recently set some words from the poetry my pupils had been learning, and they had to give the meaning, and also make a sentence for each with the word in it. The results were, in some cases, by no means creditable to the master; but I am far indeed from having attained my own ideal in this matter, or in any other. The word "*flank*" was by several said to mean the *back*. Some said a *holster* was a pistol, some that a *peer* was a man without an equal, and worst of all, not a few who had learnt the line

" The sheen of their spears was like stars on the sea,"

thought that the *sheen* was the *handle*. I believe we very few of us have any notion how small the working vocabulary of the young is; and the words outside this working vocabulary they will not trouble themselves to understand, unless their attention is specially called to them. For this reason, as well as others, we should make them thoroughly familiar with the exact meaning of all the words in their store of poetry, and we should take care that each word should suggest the line in which it occurs. A few minutes in the daily poetry lesson may be spent in ask·

ing such questions with reference to poetry already stored as, "Where does the word 'cohorts' occur?" "In what line is the 'Sea of Galilee' mentioned?" "In what way is a 'girth' mentioned in *The Ride to Aix*?" "What instance can you give of the use of the word 'bayonets?'"

I have gone into detail in this matter, because I thought that I could in this way best show you how our theory or conception of our task will make itself felt in our practice, *i.e.*, in our method of working.* But these details are, in them-

---

* I lately had a visit from a friend who is a schoolmaster, heartily interested in his profession. He wished to see my boys at work; and when he went into the school-room, he found them writing poetry from memory. Some of them were sitting biting their pens and quite aground. My friend went to these boys and asked, "Why do you stop?" "I can't remember what comes next, sir." "How do you try to remember?" This was a puzzling question. It seemed that some boys sat hopelessly trying to think of the next word, though with small prospect of doing so. Some kept saying the part they knew to themselves, in the hope that their mind would, so to speak, acquire velocity enough to carry them over the sticking-point. Others tried to think of the subject, and what was wanted to continue it at the point of difficulty. These investigations proved very interesting to both of us, and I wondered very much that I had never made them before. My friend went on to inquire *how* the boys learnt their poetry. I had talked this matter over with them, and had, as far as precepts went, put them on what I considered the right way of learning; but I found from their answers, and from a letter I got each boy to write afterwards on the subject, that these boys though intelligent and no longer children, made more use of the sound than of the sense in learning by heart. The natural divisions of the *subject* were little thought of. We do not as a rule inquire as we should *how* the work is done; and, intent on examining results, we do not observe the process by which our pupils' minds have reached them. But if we would remove our centre of interest from our own minds to the minds of

selves, of very small importance. The great thing for us to bear in mind is that we are superintending the development of our children's powers, and must subordinate all details to this central truth.   In ordinary school-life, when our energy and temper barely last out to the end of our day's work, we are too apt to lose sight of "theory" altogether, and to content ourselves with a kind of "practice" which will hardly bear thinking of.   We have, perhaps, a half-consciousness of this, and turn to what we consider necessary relaxation as soon as possible.   But there is little chance of improvement, if we settle down into a routine of this kind.   In my opinion, a teacher is wasting most valuable opportunities, if he or she does not carefully note down, in private, what the various school exercises ought to do; where they seem to fail; how they may be improved.   These private notes are almost necessary to give a continuity to our efforts, as well as to hoard our experiences.   If teachers were in the habit of rendering to themselves an account of their work, and keeping a written record for their own eyes only, much of the wretched parrot-learning of the shool-room would soon cease, and there would be far less danger than heretofore of what Mr. Brudenell Carter has too justly called *the artificial production of stupidity in schools.*

our pupils, and observe these at work, we should become better judges of results and should gain increased power of improving them.

Teachers' Manuals Series.

Each is printed in large, clear type, on good paper. Paper cover, price 15 cents; *to teachers*, 12 cents; by mail, 1 cent extra.

There is a need of small volumes—"Educational tracts," that teachers can carry easily and study as they have opportunity. The following numbers have been already published.

It should be noted that while our editions of such of these little books that are not written specially for this series are as low in price as any other, the side-heads, topics, and analyses inserted by the editor, as well as the excellent paper and printing, make them far superior in every way to any other edition.

J. G. FITCH, Inspector of the Training Colleges of England.

We would suggest that city superintendents or conductors of institutes supply each of their teachers with copies of these little books. Special rates for quantities.

No. 1.   Fitch's Art of Questioning.

By J. G. FITCH, M.A., author of "Lectures on Teaching." 38 pp.
Already widely known as the most useful and practical essay on this most important part of the teachers' lesson-hearing.

No. 2.   Fitch's Art of Securing Attention.

By J. G. FITCH, M. A. 39 pp.
Of no less value than the author's "Art of Questioning."

No. 3.   Sidgwick's On Stimulus in School.

By ARTHUR SIDGWICK, M.A. 43 pp.
"How can that dull, lazy scholar be pressed on to work up his lessons with a will?" This bright essay will tell how it can be done.

No. 4.   Yonge's Practical Work in School.

By CHARLOTTE M. YONGE, author of "Heir of Redclyffe," 35 pp.
All who have read Miss Yonge's books will be glad to read of her views on School Work.

No. 5.   Fitch's Improvement in the Art of Teaching.

By J. G. FITCH, M.A. 25 pp.
This thoughtful, earnest essay will bring courage and help to many a teacher who is struggling to do better work. It includes a course of study for Teachers' Training Classes.

No. 6. Gladstone's Object Teaching.
By J. H. GLADSTONE, of the London (Eng.) School Board. 25 pp.
A short manual full of practical suggestions on Object Teaching.

No. 7. Huntington's Unconscious Tuition.
Bishop Huntington has placed all teachers under profound obligations to him by writing this work. The earnest teacher has felt its earnest spirit, due to its interesting discussion of the foundation principles of education. It is wonderfully suggestive.

No. 8. Hughes' How to Keep Order.
By JAMES L. HUGHES, author of "Mistakes in Teaching."
Mr. Hughes is one of the few men who know what to say to help a young teacher. Thousands are to-day asking, "How shall we keep order?" Thousands are saying, "I can teach well enough, but I cannot keep order." To such we recommend this little book.

No. 9. Quick's How to Train the Memory.
By Rev. R. H. QUICK, author of "Educational Reformers."
This book comes from school-room experience, and is not a matter of theory. Much attention has been lately paid to increasing the power of memory. The teacher must make it part of his business to store the memory, hence he must know how to do it properly and according to the laws of the mind.

No. 10. Hoffman's Kindergarten Gifts.
By HEINRICH HOFFMAN, a pupil of Froebel.
The author sets forth very clearly the best methods of using them for training the child's senses and power of observation.

No. 11. Butler's Argument for Manual Training.
By NICHOLAS MURRAY BUTLER, Pres. of N. Y. College for Training of Teachers.
A clear statement of the foundation principles of Industrial Education.

No. 12. Groff's School Hygiene.
By Pres. G. G. GROFF, of Bucknell University, Pa.
We wish that every teacher could read carefully and put in practice the clearly-stated principles of School Hygiene given in this little book. Care of the eyes, light, ventilation, wells, water-closets, etc., are fully treated, with several illustrations.

THIS LIST IS CONSTANTLY BEING ADDED TO.

NOTICES.
Central School Journal (Iowa.—"The demand is for small books on great subjects."
S. W. Journal of Education.—"Glad to see such valuable papers in such a cheap form."
Va. School Journal.—"Teachers' manuals in the broad sense."
Wisconsin School Journal.—"The series are deserving the highest commendation."
Education (Boston).—"Capital little books."
Science (N. Y. City).—"Contain materials that will prove suggestive to teachers."
Progressive Teacher.—"Valuable additions to a series already famous."
School Herald (Chicago).—"We must commend the good judgment in selecting these books."
Educational Record (Canada).—"Every progressive teacher ought to have them."

Song Treasures.

THE PRICE HAS BEEN GREATLY REDUCED.

Compiled by AMOS M. KELLOGG, editor of the SCHOOL JOUR-NAL. Beautiful and durable postal-card manilla cover, printed in two colors, 64 pp. Price, 15 cents each; *to teachers,* 12 cents; by mail, 2 cents extra. 30th thousand. *Write for our special terms to schools for quantities. Special terms for use at Teachers' Institutes.*

This is a most valuable collection of music for all schools and institutes.

1. Most of the pieces have been selected by the teachers as favorites in the schools. They are the ones the pupils love to sing. It contains nearly 100 pieces.

2. All the pieces "have a ring to them;" they are easily learned, and will not be forgotten.

3. The themes and words are appropriate for young people. In these respects the work will be found to possess unusual merit. Nature, the Flowers, the Seasons, the Home, our Duties, our Creator, are entuned with beautiful music.

4. Great ideas may find an entrance into the mind through music. Aspirations for the good, the beautiful, and the true are presented here in a musical form.

5. Many of the words have been written especially for the book. One piece, "The Voice Within Us," p. 57, is worth the price of the book.

6. The titles here given show the teacher what we mean:

Ask the Children, Beauty Everywhere, Be in Time, Cheerfulness, Christmas Bells, Days of Summer Glory, The Dearest Spot, Evening Song, Gentle Words, Going to School, Hold up the Right Hand, I Love the Merry, Merry Sunshine, Kind Deeds, Over in the Meadows, Our Happy School, Scatter the Germs of the Beautiful, Time to Walk, The Jolly Workers, The Teacher's Life, Tribute to Whittier, etc., etc.

Welch's Teachers' Psychology.

A Treatise on the Intellectual Faculties, the Order of the Growth, and the Corresponding Series of Studies by which they are Educated. By the late A. S. Welch, Professor of Psychology, Iowa Agricultural College, formerly Pres. of the Mich. Normal School. Cloth, 12mo, 300 pp., $1.25; *to teachers*, $1; by mail, 12 cents extra. Special terms to Normal Schools and Reading Circles.

A mastery of the branches to be taught was once thought to be an all-sufficient preparation for teaching. But it is now seen that there must be a knowledge of the mind that is to be trained. Psychology is the foundation of intelligent pedagogy. Prof. Welch undertook to write a book that should deal with mind-unfolding, as exhibited in the school-room. He shows what is meant by attending, memorizing, judging, abstracting, imagining, classifying, etc., as it is done by the pupil over his text-books. First, there is the *concept;* then there is (1) gathering concepts, (2) storing concepts, (3) dividing concepts, (4) abstracting concepts, (5) building concepts, (6) grouping concepts, (7) connecting concepts, (8) deriving concepts. Each of these is clearly explained and illustrated; the reader instead of being bewildered over strange terms comprehends that imagination means a building up of concepts, and so of the other terms.

DR. A. S. WELCH.

A most valuable part of the book is its application to practical education. How to train these powers that deal with the concept—that is the question. There must be exercises to train the mind to *gather, store, divide, abstract, build, group, connect,* and *derive* concepts. The author shows what studies do this appropriately, and where there are mistakes made in the selection of studies. The book will prove a valuable one to the teacher who wishes to know the structure of the mind and the way to minister to its growth. It would seem that at last a psychology had been written that would be a real aid, instead of a hindrance, to clear knowledge.

As a text-book for the use of students in normal schools, teachers' institutes, reading circles, etc., this book is unsurpassed. The logical arrangement, the directness of presentation, without unnecessary words or repetition, the questions at end of each chapter, and typographical features, make it an ideal text-book. Only two months after publication it was introduced into many normal schools as a text-book, and adopted by the Cal. State Teachers' Reading Circle.

OUTLINE OF CONTENTS.

Psychology.

Psychology and Education.

This book is written by one who, as a teacher, institute conductor, president of a normal school (Mich., 15 years), president of college (Iowa, for many years), has shown himself to be a thoughtful student of education. He has made the volume one that will *aid the teacher in carrying forward the school-room work in accordance with mind laws.* So great has been the interest created that 1,000 COPIES WERE ORDERED IN ADVANCE of publication. Dr. Welch's book is a large 12mo volume of 300 pp. beautifully printed from large, clear type, and artistically and durably bound. As so many teachers are making inquiries on psychological points, we feel certain that they will find this book just what they want.

Reception Day. 6 Nos.

A collection of fresh and original dialogues, recitations, declamations, and short pieces for practical use in Public and Private Schools. Bound in handsome new paper cover, 100 pages each, printed on laid paper. Price, 30 cents each; *to teachers*, 24 cents; by mail, 3 cents extra.

The exercises in these books bear upon education; have a relation to the school-room.

NEW COVER.

1. The dialogues, recitations, and declamations gathered in this volume being fresh, short, and easy to be comprehended, are well fitted for the average scholars of our schools.

2. They have mainly been used by teachers for actual school exercises.

3. They cover a different ground from the speeches of Demosthenes and Cicero—which are unfitted for boys of twelve to sixteen years of age.

4. They have some practical interest for those who use them.

5. There is not a vicious sentence uttered. In some dialogue books profanity is found, or disobedience to parents encouraged, or lying laughed at. Let teachers look out for this.

6. There is something for the youngest pupils.

7. "Memorial Day Exercises" for Bryant, Garfield, Lincoln, etc., will be found.

8. Several Tree Planting exercises are included.

9. The exercises have relation to the school-room, and bear upon education.

10. An important point is the freshness of these pieces. Most of them were written expressly for this collection, and *can be found nowhere else.*

Boston Journal of Education.—"It is of practical value."
Detroit Free Press.—"Suitable for public and private schools."
Western Ed. Journal.—"A series of very good selections."

WHAT EACH NUMBER CONTAINS.

No. 1

Is a specially fine number. One dialogue in it, called "Work Conquers," for 11 girls and 6 boys, has been given hundreds of times, and is alone worth the price of the book. Then there are 21 other dialogues.
29 Recitations.
14 Declamations.
17 Pieces for the Primary Class.

No. 2 Contains

29 Recitations.
12 Declamations.
17 Dialogues.
24 Pieces for the Primary Class.
And for Class Exercise as follows:
The Bird's Party.
Indian Names.
Valedictory.
Washington's Birthday.
Garfield Memorial Day.
Grant   "   "
Whittier   "   "
Sigourney "   "

No. 3 Contains

Fewer of the longer pieces and more of the shorter, as follows:
18 Declamations.
21 Recitations.
22 Dialogues.
24 Pieces for the Primary Class.
A Christmas Exercise.
Opening Piece, and
An Historical Celebration.

No. 4 Contains

Campbell Memorial Day.
Longfellow   "   "
Michael Angelo "   "
Shakespeare   "   "
Washington   "   "
Christmas Exercise.
Arbor Day   "
New Planting "
Thanksgiving "
Value of Knowledge Exercise.
Also 8 other Dialogues.
21 Recitations.
23 Declamations.

No. 5 Contains

Browning Memorial Day.
Autumn Exercise.
Bryant Memorial Day.
New Planting Exercise.
Christmas Exercise.
A Concert Exercise.
24 Other Dialogues.
16 Declamations, and
36 Recitations.

No. 6 Contains

Spring; a flower exercise for very young pupils.
Emerson Memorial Day.
New Year's Day Exercise.
Holmes' Memorial Day.
Fourth of July Exercise.
Shakespeare Memorial Day.
Washington's Birthday Exercise.
Also 6 other Dialogues.
6 Declamations.
41 Recitations.
15 Recitations for the Primary Class.
And 4 Songs.

Our RECEPTION DAY Series is not sold largely by booksellers, who, if they do not keep it, try to have you buy something else similar, but not so good. Therefore send direct to the publishers, by mail, the price as above, in stamps or postal notes, and your order will be filled at once. Discount for quantities.

SPECIAL OFFER.

If ordered at one time, we will send postpaid the entire 6 Nos. for $1.40. Note the reduction.

Perez's First Three Years of Childhood.

An Exhaustive Study of the Psychology of Children. By BERNARD PEREZ. Edited and translated by ALICE M. CHRISTIE, translator of "Child and Child Nature," with an introduction by JAMES SULLY, M.A., author of "Outlines of Psychology," etc. 12mo, cloth, 324 pp. Price, $1.50; *to teachers*, $1.20; by mail, 10 cents extra.

This is a comprehensive treatise on the psychology of childhood, and is a practical study of the human mind, not full formed and equipped with knowledge, but as nearly as possible, *ab origine*—before habit, environment, and education have asserted their sway and made their permanent modifications. The writer looks into all the phases of child activity. He treats exhaustively, and in bright Gallic style, of sensations, instincts, sentiments, intellectual tendencies, the will, the faculties of æsthetic and moral senses of young children. He shows how ideas of truth and falsehood arise in little minds, how natural is imitation and how deep is credulity. He illustrates the development of imagination and the elaboration of new concepts through judgment, abstraction, reasoning, and other mental methods. It is a book that has been long wanted by all who are engaged in teaching, and especially by all who have to do with the education and training of children.

This edition has a new index of special value, and the book is carefully printed and elegantly and durably bound. Be sure to get our standard edition.

OUTLINE OF CONTENTS.

CHAP.
I. Faculties of Infant before Birth —First Impression of Newborn Child.
II. Motor Activity at the Beginning of Life—at Six Months— —at Fifteen Months.
III. Instinctive and Emotional Sensations—First Perceptions.
IV. General and Special Instincts.
V. The Sentiments.
VI. Intellectual Tendencies—Veracity—Imitation—Credulity.
VII. The Will.
VIII. Faculties of Intellectual Acquisition and Retention—Attention—Memory.

CHAP.
IX. Association of Psychical States – Association—Imagination.
X. Elaboration of Ideas—Judgment — Abstraction — Comparison — Generalization — Reasoning—Errors and Allusions—Errors and Allusions Owing to Moral Causes.
XI. Expression and Language.
XII. Æsthetic Senses — Musical Sense — Sense of Material Beauty — Constructive Instinct—Dramatic Instinct.
XIII. Personalty — Reflection—Moral Sense.

Col. Francis W. Parker, Principal Cook County Normal and Training School, Chicago, says:—"I am glad to see that you have published Perez's wonderful work upon childhood. I shall do all I can to get everybody to read it. It is a grand work."

John Bascom, Pres. Univ. of Wisconsin, says:—"A work of marked interest."

G. Stanley Hall, Professor of Psychology and Pedagogy, Johns Hopkins Univ., says:—"I esteem the work a very valuable one for primary and kindergarten teachers, and for all interested in the psychology of childhood."
And many other strong commendations.

Allen's Mind Studies for Young Teach-

ERS. By JEROME ALLEN, Ph.D., Associate Editor of the SCHOOL JOURNAL, Prof. of Pedagogy, Univ. of City of N. Y. 16mo, large, clear type, 128 pp. Cloth, 50 cents; *to teachers*, 40 cents; by mail, 5 cents extra.

JEROME ALLEN, Ph.D., Associate Editor of the *Journal* and *Institute*.

There are many teachers who know little about psychology, and who desire to be better informed concerning its principles, especially its relation to the work of teaching. For the aid of such, this book has been prepared. But it is not a psychology—only an introduction to it, aiming to give some fundamental principles, together with something concerning the philosophy of education. Its method is subjective rather than objective, leading the student to watch mental processes, and draw his own conclusions. It is written in language easy to be comprehended, and has many practical illustrations. It will aid the teacher in his daily work in dealing with mental facts and states.

To most teachers psychology seems to be dry. This book shows how it may become the most interesting of all studies. It also shows how to begin the knowledge of self. "We cannot know in others what we do not first know in ourselves." This is the key-note of this book. Students of elementary psychology will appreciate this feature of "Mind Studies."

ITS CONTENTS.

CHAP.
I. How to Study Mind.
II. Some Facts in Mind Growth.
III. Development.
IV. Mind Incentives.
V. A few Fundamental Principles Settled.
VI. Temperaments.
VII. Training of the Senses.
VIII. Attention.
IX. Perception.
X. Abstraction.
XI. Faculties used in Abstract Thinking.

CHAP.
XII. From the Subjective to the Conceptive.
XIII. The Will.
XIV. Diseases of the Will.
XV. Kinds of Memory.
XVI. The Sensibilities.
XVII. Relation of the Sensibilities to the Will.
XVIII. Training of the Sensibilities.
XIX. Relation of the Sensibilities to Morality.
XX. The Imagination.
XXI. Imagination in its Maturity.
XXII. Education of the Moral Sense.

Browning's Educational Theories.

By OSCAR BROWNING, M.A., of King's College, Cambridge, Eng. No. 8 of *Reading Circle Library Series.* Cloth, 16mo, 237 pp. Price, 50 cents; *to teachers,* 40 cents; by mail, 5 cents extra.

This work has been before the public some time, and for a general sketch of the History of Education it has no superior. Our edition contains several new features, making it specially valuable as a text-book for Normal Schools, Teachers' Classes, Reading Circles, Teachers' Institutes, etc., as well as the student of education. These new features are: (1) Side-heads giving the subject of each paragraph; (2) each chapter is followed by an analysis; (3) a very full *new* index; (4) also an appendix on "Froebel," and the "American Common School."

OUTLINE OF CONTENTS.

I. Education among the Greeks—Music and Gymnastic Theories of Plato and Aristotle; II. Roman Education—Oratory; III. Humanistic Education; IV. The Realists—Ratich and Comenius; V. The Naturalists—Rabelais and Montaigne; VI. English Humorists and Realists—Roger Ascham and John Milton; VII. Locke; VIII. Jesuits and Jansenists; IX. Rousseau; X. Pestalozzi; XI. Kant, Fichte, and Herbart; XII. The English Public School; XIII. Froebel; XIV. The American Common School.

PRESS NOTICES.

Ed. Courant.—"This edition surpasses others in its adaptability to general use."

Col. School Journal.—"Can be used as a text-book in the History of Education."

Pa. Ed. News.—"A volume that can be used as a text-book on the History of Education."

School Education, Minn.—"Beginning with the Greeks, the author presents a brief but clear outline of the leading educational theories down to the present time."

Ed. Review, Can.—"A book like this, introducing the teacher to the great minds that have worked in the same field, cannot but be a powerful stimulus to him in his work."

Dewey's How to Teach Manners in the

SCHOOL-ROOM. By Mrs. JULIA M. DEWEY, Principal of the Normal School at Lowell, Mass., formerly Supt. of Schools at Hoosick Falls, N. Y. Cloth, 16mo, 104 pp. Price, 50 cents; *to teachers*, 40 cents; by mail, 5 cents extra.

Many teachers consider the manners of a pupil of little importance so long as he is industrious. But the boys and girls are to be fathers and mothers; some of the boys will stand in places of importance as professional men, and they will carry the mark of ill-breeding all their lives. Manners can be taught in the school-room: they render the school-room more attractive; they banish tendencies to misbehavior. In this volume Mrs. Dewey has shown how manners can be taught. The method is to present some fact of deportment, and then lead the children to discuss its bearings; thus they learn why good manners are to be learned and practised. The printing and binding are exceedingly neat and attractive."

OUTLINE OF CONTENTS.

Central School Journal.—"It furnishes illustrative lessons."

Texas School Journal.—"They (the pupils) will carry the mark of ill breeding all their lives (unless taught otherwise)."

Pacific Ed. Journal.—"Principles are enforced by anecdote and conversation."

Teacher's Exponent.—"We believe such a book will be very welcome."

National Educator.—"Common-sense suggestions."

Ohio Ed. Monthly.—"Teachers would do well to get it."

Nebraska Teacher.—"Many teachers consider manners of little importance, but some of the boys will stand in places of importance."

School Educator.—"The spirit of the author is commendable."

School Herald.—"These lessons are full of suggestions."

Va. School Journal.—"Lessons furnished in a delightful style."

Miss. Teacher.—"The best presentation we have seen."

Ed. Courant.—"It is simple, straightforward, and plain."

Iowa Normal Monthly.—"Practical and well-arranged lessons on manners."

Progressive Educator.—"Will prove to be most helpful to the teacher who desires her pupils to be well-mannered."

Fitch's Lectures on Teaching.

Lectures on Teaching. By J. G. FITCH, M.A., one of Her
Majesty's Inspectors of Schools. England.   Cloth, 16mo,
395 pp. Price, $1.25 ; *to teachers*, $1.00 ; by mail, postpaid.

Mr. Fitch takes as his topic the application of principles to
the art of teaching in schools.   Here are no vague and general propositions, but on every page we find the problems of
the school-room discussed with definiteness of mental grip.
No one who has read a single lecture by this eminent man
but will desire to read another.   The book is full of suggestions that lead to increased power.

1. These lectures are highly prized in England.
2. There is a valuable preface by Thos. Hunter, President
of N. Y. City Normal College.
3. The volume has been at once adopted by several State
Reading Circles.

EXTRACT FROM AMERICAN PREFACE.

"Teachers everywhere among English-speaking people have hailed
Mr. Fitch's work as an invaluable aid for almost every kind of instruction and school organization. It combines the theoretical and the practical; it is based on psychology; it gives admirable advice on everything connected with teaching—from the furnishing of a school-room
to the preparation of questions for examination. Its style is singularly
clear, vigorous and harmonious."

Chicago Intelligence.—"All of its discussions are based on sound
psychological principles and give admirable advice."

Virginia Educational Journal.—"He tells what he thinks so as to
be helpful to all who are striving to improve."

Lynn Evening Item.—"He gives admirable advice."

Philadelphia Record.—"It is not easy to imagine a more useful volume."

Wilmington Every Evening.—"The teacher will find in it a wealth
of help and suggestion."

Brooklyn Journal.—"His conception of the teacher is a worthy ideal
for all to bear in mind."

New England Journal of Education : "This is eminently the work of
a man of wisdom and experience. He takes a broad and comprehensive
view of the work of the teacher, and his suggestions on all topics are
worthy of the most careful consideration."

Brooklyn Eagle : "An invaluable aid for almost every kind of instruction and school organization. It combines the theoretical and the
practical; it is based on psychology; it gives admirable advice on everything connected with teaching, from the furnishing of a school-room to
the preparation of questions for examination."

Toledo Blade : "It is safe to say, no teacher can lay claim to being
well informed who has not read this admirable work. Its appreciation
is shown by its adoption by several State Teachers' Reading Circles, as
a work to be thoroughly read by its members."

Hughes' Mistakes in Teaching.

By James J. Hughes, Inspector of Schools, Toronto, Canada. Cloth, 16mo, 115 pp. Price, 50 cents; *to teachers*, 40 cents; by mail, 5 cents extra.

Thousands of copies of the old edition have been sold. The new edition is worth double the old; the material has been increased, restated, and greatly improved. Two new and important Chapters have been added on "Mistakes in Aims," and "Mistakes in Moral Training." Mr. Hughes says in his preface: "In issuing a revised edition of this book, it seems fitting to acknowledge gratefully the hearty appreciation that has been accorded it by American teachers. Realizing as I do that its very large sale indicates that it has been of service to many of my fellow-teachers, I have recognized the duty of enlarging and revising it so as to make it still more helpful in preventing the common mistakes in teaching and training."

JAMES L. HUGHES, Inspector of Schools, Toronto, Canada.

This is one of the six books recommended by the N. Y. State Department to teachers preparing for examination for State certificates.

CAUTION.

Our new AUTHORIZED COPYRIGHT EDITION, *entirely rewritten by the author, is the only one to buy. It is beautifully printed and handsomely bound. Get no other.*

CONTENTS OF OUR NEW EDITION.

CHAP. I. 7 Mistakes in Aim.

CHAP. II. 21 Mistakes in School Management.

CHAP. III. 24 Mistakes in Discipline.

CHAP. IV. 27 Mistakes in Method.

CHAP. V. 13 Mistakes in Moral Training.

☞ *Chaps. I. and V. are entirely new.*

Hughes' Securing and Retaining Atten-

TION. By JAMES L. HUGHES, Inspector Schools, Toronto, Canada, author of "Mistakes in Teaching." Cloth, 116 pp. Price, 50 cents; *to teachers*, 40 cents; by mail, 5 cents extra.

This valuable little book has already become widely known to American teachers. Our new edition has been almost *entirely re-written*, and several new important chapters added. It is the only AUTHORIZED COPYRIGHT EDITION. *Caution.*—Buy no other.

WHAT IT CONTAINS.

I. General Principles; II. Kinds of Attention; III. Characteristics of Good Attention; IV. Conditions of Attention; V. Essential Characteristics of the Teacher in Securing and Retaining Attention; VI. How to Control a Class; VII. Methods of Stimulating and Controlling a Desire for Knowledge; VIII. How to Gratify and Develop the Desire for Mental Activity; IX. Distracting Attention; X. Training the Power of Attention; XI. General Suggestions regarding Attention.

TESTIMONIALS.

S. P. Robbins, Pres. McGill Normal School, Montreal, Can., writes to Mr. Hughes:—"It is quite superfluous for me to say that your little books are admirable. I was yesterday authorized to put the 'Attention' on the list of books to be used in the Normal School next year. Crisp and attractive in style, and mighty by reason of its good, sound common-sense, it is a book that every teacher should know."

Popular Educator (Boston):—"Mr. Hughes has embodied the best thinking of his life in these pages."

Central School Journal (Ia.).—"Though published four or five years since, this book has steadily advanced in popularity."

Educational Courant (Ky.).—"It is intensely practical. There isn't a mystical, muddy expression in the book."

Educational Times (England).—"On an important subject, and admirably executed."

School Guardian (England).—"We unhesitatingly recommend it."

New England Journal of Education.—"The book is a guide and a manual of special value."

New York School Journal.—"Every teacher would derive benefit from reading this volume."

Chicago Educational Weekly.—"The teacher who aims at best success should study it."

Phil. Teacher.—"Many who have spent months in the school-room would be benefited by it."

Maryland School Journal.—"Always clear, never tedious."
Va. Ed. Journal.—"Excellent hints as to securing attention."
Ohio Educational Monthly.—"We advise readers to send for a copy."
Pacific Home and School Journal.—"An excellent little manual."
Prest. James H. Hoose, State Normal School, Cortland, N. Y., says:— "The book must prove of great benefit to the profession."
Supt. A. W. Edson, Jersey City, N. J., says:—"A good treatise has long been needed, and Mr. Hughes has supplied the want."

Payne's Lectures on the Science and

ART OF EDUCATION. *Reading Circle Edition.* By JOSEPH PAYNE, the first Professor of the Science and Art of Education in the College of Preceptors, London, England. With portrait. 16mo, 350 pp., English cloth, with gold back stamp. Price, $1.00 ; *to teachers,* 80 cents ; by mail, 7 cents extra. *Elegant new edition from new plates.*

JOSEPH PAYNE.

Teachers who are seeking to know the principles of education will find them clearly set forth in this volume. It must be remembered that principles are the basis upon which all methods of teaching must be founded. So valuable is this book that if a teacher were to decide to own but three works on education, this would be one of them. This edition contains all of Mr. Payne's writings that are in any other American abridged edition, and *is the only one with his portrait.* It is far superior to any other edition published.

WHY THIS EDITION IS THE BEST.

(1.) The *side-titles.* These give the contents of the page. (2.) The analysis of each lecture, with reference to the *educational* points in it. (3.) The general analysis pointing out the three great principles found at the beginning. (4.) The index, where, under such heads as Teaching, Education, The Child, the important utterances of Mr. Payne are set forth. (5.) Its handy shape, large type, fine paper, and press-work and tasteful binding. All of these features make this a most valuable book. To obtain all these features in one edition, it was found necessary to *get out this new edition.*

Ohio Educational Monthly.—"It does not deal with shadowy theories; it is intensely practical."

Philadelphia Educational News.—"Ought to be in library of every progressive teacher."

Educational Courant.—"To know how to teach, more is needed than a knowledge of the branches taught. This is especially valuable."

Pennsylvania Journal of Education.—"Will be of practical value to Normal Schools and Institutes

Parker's Talks on Teaching.

Notes of "Talks on Teaching" given by COL. FRANCIS W. PARKER (formerly Superintendent of schools of Quincy, Mass.), before the Martha's Vineyard Institute, Summer of 1882. Reported by LELIA E. PATRIDGE. Square 16mo, 5x6 1-2 inches, 192 pp., *laid* paper, English cloth. Price, $1.25 ; *to teachers,* $1.00 ; by mail, 9 cents extra.

The methods of teaching employed in the schools of Quincy, Mass., were seen to be the methods of nature. As they were copied and explained, they awoke a great desire on the part of those who could not visit the schools to know the underlying principles. In other words, Colonel Parker was asked to explain *why* he had his teachers teach thus. In the summer of 1882, in response to requests, Colonel Parker gave a course of lectures before the Martha's Vineyard Institute, and these were reported by Miss Patridge, and published in this book.

The book became famous ; more copies were sold of it in the same time than of any other educational book whatever. The daily papers, which usually pass by such books with a mere mention, devoted columns to reviews of it.

The following points will show why the teacher will want this book.

1. It explains the "New Methods." There is a wide gulf between the new and the old education. Even school boards understand this.

2. It gives the underlying principles of education. For it must be remembered that Col. Parker is not expounding *his* methods, but the methods of nature.

3. It gives the ideas of a man who is evidently an "educational genius," a man born to understand and expound education. We have few such ; they are worth everything to the human race.

4. It gives a biography of Col. Parker. This will help the teacher of education to comprehend the man and his motives.

5. It has been adopted by nearly every State Reading Circle

Patridge's " Quincy Methods."

The " Quincy Methods," illustrated ; Pen photographs from the Quincy schools. By LELIA E. PATRIDGE. Illustrated with a number of engravings, and two colored plates. Blue cloth, gilt, 12mo, 686 pp. Price, $1.75 ; *to teachers,* $1.40 ; by mail, 13 cents extra.

When the schools of Quincy, Mass., became so famous under the superintendence of Col. Francis W. Parker, thousands of teachers visited them. Quincy became a sort of " educational Mecca," to the disgust of the routinists, whose schools were passed by. Those who went to study the methods pursued there were called on to tell what they had seen. Miss Patridge was one of those who visited the schools of Quincy ; in the Pennsylvania Institutes (many of which she conducted), she found the teachers were never tired of being told how things were done in Quincy. She revisited the schools several times, and wrote down what she saw ; then the book was made.

1. This book presents the actual practice in the schools of Quincy. It is composed of " pen photographs."

2. It gives abundant reasons for the great stir produced by the two words " Quincy Methods." There are reasons for the discussion that has been going on among the teachers of late years.

3. It gives an insight to principles underlying real education as distinguished from book learning.

4. It shows the teacher not only what to do, but gives the way in which to do it.

5. It impresses one with the *spirit* of the Quincy schools.

6. It shows the teacher how to create an *atmosphere* of happiness, of busy work, and of progress.

7. It shows the teacher how not to waste her time in worrying over disorder.

8. It tells how to treat pupils with courtesy, and get courtesy back again.

9. It presents four years of work, considering Number, Color, Direction, Dimension, Botany, Minerals, Form, Language, Writing, Pictures, Modelling, Drawing, Singing, Geography, Zoology, etc., etc.

10. There are 686 pages; a large book devoted to the realities of school life, in realistic descriptive language. It is plain, real, not abstruse and uninteresting.

11. It gives an insight into real education, the education urged by Pestalozzi, Frœbel, Mann, Page, Parker, etc.

Woodbull's Simple Experiments for the

SCHOOL-ROOM. By Prof. JOHN F. WOODHULL, Prof. of Natural Science in the College for the Training of Teachers, New York City, author of "Manual of Home-Made Apparatus." Cloth, 16mo. Price, 50 cents; *to teachers,* 40 cents; by mail, 5 cents extra.

This book contains a series of simple, easily-made experiments, to perform which will aid the comprehension of every-day phenomena. They are really the very lessons given by the author in the Primary and Grammar Departments of the Model School in the College for the Training of Teachers, New York City.

The apparatus needed for the experiments consists, for the most part, of such things as every teacher will find at hand in a school-room or kitchen. The experiments are so connected in logical order as to form a continuous exhibition of the phenomena of *combustion. This book is not a science catechism.* Its aim is to train the child's mind in habits of reasoning by experimental methods.

These experiments should be made in every school of our country, and thus bring in a scientific method of dealing with nature. The present method of cramming children's minds with isolated facts of which they can have no adequate comprehension is a ruinous and unprofitable one. This book points out the method employed by the *best teachers in the best schools.*

WHAT IT CONTAINS.

I. Experiments with Paper.	VI.	Air as an Agent in Combustion.
II. " " Wood.	VII.	Products of Complete "
III. " " a Candle.	VIII.	Currents of Air, etc.—Ventila-
IV. " " Kerosene.	IX.	Oxygen of the Air. [tion.
V. Kindling Temperature.	X.	Chemical Changes.

In all there are 91 experiments described, illustrated by 35 engravings.

Jas. H. Canfield, Univ. of Kans., Lawrence, says:—"I desire to say most emphatically that the method pursued is the only true one in all school work. Its spirit is admirable. We need and must have far more of this instruction."

J. C. Packard, Univ. of Iowa, Iowa City, says:—"For many years shut up to the simplest forms of illustrative apparatus, I learned that the necessity was a blessing, since so much could be accomplished by home-made apparatus—inexpensive and effective."

Henry R. Russell, Woodbury, N. J., Supt. of the Friends School:—"Admirable little book. It is just the kind of book we need."

S. T. Dutton, Supt. Schools, New Haven, Ct.—"Contains just the kind of help teachers need in adapting natural science to common schools."

Wilhelm's Student's Calendar.

Compiled by N. O. WILHELM. Bound in paper. 76 pp. Double indexed. Price, 30 cents; *to teachers*, 24 cents; by mail, 3 cents extra.

This is a perpetual calendar and book of days. It consists of Short Biographies of Greatest Men, arranged according to Birthdays and Deathdays, covering every day of the year.

These can be used for opening exercises in schools, for memorial days, and for giving pupils some information about the great men of the world about whom everybody ought to know something. Just the thing for families where there are young people.

The condensed information in this little book would in other form cost you many dollars to own. Here are a few of the names of persons whose Biographies are found in the "Student's Calendar:"

John Adams,	Queen Elizabeth,	John Hancock,	Abraham Lincoln,
J. Q. Adams,	R. W. Emerson,	Hamilton,	Jenny Lind,
Joseph Addison,	Robert Emmet,	Hannibal,	Linnæus,
Alexand'r the Gre't,	Euripides,	W. H. Harrison,	Dr. Livingstone,
Michael Angelo,	Edw. Everett,	Nath. Hawthorne,	H. W. Longfellow,
Aristotle,	Faraday,	Hayden,	Lowell,
Ascham,	Farragut,	Mrs. Hemans,	Lubbock,
Audubon,	Fénelon,	T. A. Hendricks,	Martin Luther,
Francis Bacon,	M. Fillmore,	Patrick Henry,	Macaulay,
Geo. Bancroft,	Chas. J. Fox,	Sir Wm. Herschel,	Macready,
Venerable Bede,	Ben. Franklin,	O. W. Holmes,	Mohammed,
Von Bismarck,	Sir J. Franklin,	Thomas Hood,	Horace Mann,
Tycho Brahe,	Frederick the Great	Jos. Hooker,	Maria Theresa,
Lord Brougham,	J. C. Fremont,	Horace,	Marie Antoinette,
Mrs. Browning,	Frobisher,	Sam. Houston,	Mary, Qu'n of Scots
W. C. Bryant,	Froebel,	Elias Howe,	J. Montgomery,
Edmund Burke,	Froude,	Victor Hugo,	Sir J. Moore,
Robert Burns,	Robert Fulton,	Humboldt,	Mozart,
Ben. F. Butler,	Galileo,	David Hume,	Napoleon I.,
Lord Byron,	Vasco da Gama,	Wash. Irving,	Nelson,
Cæsar,	Gambetta,	Andrew Jackson,	Sir Isaac Newton,
John Calhoun,	Garfield,	Jacotot,	Daniel O'Connell,
Thos. Campbell,	Garibaldi,	Jos. Jacquard,	Charles O'Conor,
Thos. Carlyle,	D. Garrick,	James I.,	Thos. Paine,
Phœbe Cary,	Horatio Gates,	James II.,	Geo. Peabody,
Cervantes,	R. Gatling,	John Jay,	Wm. Penn,
Salmon P. Chase,	George III.,	Thos. Jefferson,	Peter the Great.
Thos. Chatterton,	Stephen Girard,	Francis Jeffrey,	Pizarro,
Rufus Choate,	Gladstone,	Dr. Ed. Jenner,	Plato,
Cicero,	Goethe,	Joan of Arc,	E. A. Poe,
Henry Clay,	Goldsmith,	Sam'l Johnson,	W. H. Prescott,
Cleopatra,	U. S. Grant,	John Paul Jones,	Pulaski,
Coleridge,	Henry Grattan,	Dr. Kane,	Queen Victoria,
Schuyler Colfax,	Asa Gray,	John Keats,	Richelieu,
Anthony Collins,	Horace Greeley,	John Kitto,	J. P. Richter,
Cornwallis,	Nath. Greene,	Henry Knox,	Ritter,

Lubbock's Best 100 Books.

By Sir JOHN LUBBOCK. 64 pages, paper. Price, 20 cents; *to teachers*, 16 cents; by mail, 2 cents extra.

Sir John Lubbock, in an address last year before the Workingmen's College of London, England, gave a list of what he deemed the Best 100 Books. He said, in giving his list, that if a few good guides would draw up similar lists, it would be most useful.

The *Pall Mall Gazette* published Sir John Lubbock's list, and invited eminent men in England to give their opinions concerning it. We have just reprinted them in neat pamphlet form. Gladstone, Stanley, Black, and many others are represented.

Love's Industrial Education.

Industrial Education ; a guide to Manual Training. By SAMUEL G. LOVE, principal of the Jamestown, (N. Y.) public schools. Cloth, 12mo, 330 pp. with 40 full-page plates containing nearly 400 figures. Price, $1.50; *to teachers,* $1.20 ; by mail, 12 cents extra.

1. *Industrial Education not understood.* Probably the only man who has wrought out the problem in a practical way is

Samuel G. Love, the superintendent of the Jamestown (N. Y.) schools. Mr. Love has now about 2,400 children in the primary, advanced, and high schools under his charge ; he is assisted by fifty teachers, so that an admirable opportunity was offered. In 1874 (about fourteen years ago) Mr. Love began his experiment ; gradually he introduced one occupation, and then another, until at last nearly all the pupils are following some form of educating work.

2. *Why it is demanded.* The reasons for introducing it are clearly stated by Mr. Love. It was done because the education of the books left the pupils unfitted to meet the practical problems the world asks them to solve. The world does not have a field ready for the student in book-lore. The statements of Mr. Love should be carefully read.

3. *It is an educational book.* Any one can give some formal work to girls and boys. What has been needed has been some one who could find out what is suited to the little child who is in the " First Reader," to the one who is in the " Second Reader," and so on. It must be remembered the effort is not to make carpenters, and type-setters, and dressmakers of boys and girls, but to *educate them by these occupations better than without them.*

Seeley's Grube's Method of Teaching

ARITHMETIC. Explained and illustrated. Also the improvements on the method made by the followers of Grubé in Germany. By LEVI SEELEY, Ph.D. Cloth, 176 pp. Price, $1.00; *to teachers* 80 cents; by mail, 7 cents extra.

DR. LEVI SEELEY.

1. IT IS A PHILOSOPHICAL WORK.—This book has a sound philosophical basis. The child does not (as most teachers seem to think) learn addition, then subtraction, then multiplication, then division; he learns these processes together. Grubé saw this, and founded his system on this fact.

2. IT FOLLOWS NATURE'S PLAN.—Grubé proceeds to develop (so to speak) the method by which the child actually becomes (if he ever does) acquainted with 1, 2, 3, 4, 5, etc. This is not done, as some suppose, by writing them on a slate. Nature has her method; she begins with THINGS; after handling two things in certain ways, the idea of *two* is obtained, and so of other numbers. *The chief value of this book then consists in showing what may be termed the way nature teaches the child number.*

3. IT IS VALUABLE TO PRIMARY TEACHERS.—It begins and shows how the child can be taught 1, then 2, then 3, &c. Hence it is a work especially valuable for the primary teacher. It gives much space to showing how the numbers up to 10 are taught; for if this be correctly done, the pupil will almost teach himself the rest.

4. IT CAN BE USED IN ADVANCED GRADES.—It discusses methods of teaching fractions, percentage, etc., so that it is a work valuable for all classes of teachers.

5 IT GUIDES THE TEACHER'S WORK.—It shows, for example, what the teacher can appropriately do the first year, what the second, the third, and the fourth. More than this, it suggests work for the teacher she would otherwise omit.

Taking it altogether, it is the best work on teaching *number* ever published. It is very handsomely printed and bound

Kellogg's School Management:

" A Practical Guide for the Teacher in the School-Room."
By Amos M. Kellogg, A.M. Sixth edition. Revised and
enlarged. Cloth, 128 pp. Price, 75 cents ; *to teachers*, 60
cents ; by mail, 5 cents extra.

This book takes up the most difficult of all school work,
viz.: the Government of a school, and is filled with original
and practical ideas on the subject. It is invaluable to the
teacher who desires to make his school a " well-governed "
school.

1. It suggests methods of awakening an interest in the
studies, and in school work. "The problem for the teacher,"
says Joseph Payne, " is to get the pupil to study." If he can do
this he will be educated.

2. It suggests methods of making the school attractive.
Ninety-nine hundredths of the teachers think young people
should come to school anyhow ; the wise ones know that a
pupil who wants to come to school will do something when
he gets there, and so make the school attractive.

3. Above all it shows that the pupils will be self-governed
when well governed. It shows how to develop the process of
self-government.

4. It shows how regular attention and courteous behaviour
may be secured.

5. It has an admirable preface by that remarkable man and
teacher, Dr. Thomas Hunter, Pres. N. Y. City Normal College.

Home and School.—" Is just the book for every teacher who wishes
to be a better teacher."

Educational Journal.—" It contains many valuable hints."

Boston Journal of Education.—" It is the most humane, instructive,
original educational work we have read in many a day."

Wis. Journal of Education.—" Commends itself at once by the number of ingenious devices for securing order, industry, and interest.

Iowa Central School Journal.—"Teachers will find it a helpful and
suggestive book."

Canada Educational Monthly.—" Valuable advice and useful suggestions."

Normal Teacher.—" The author believes the way to manage is to civ-
ilize, cultivate, and refine."

School Moderator.—" Contains a large amount of valuable reading ;
school government is admirably presented."

Progressive Teacher.—"Should occupy an honored place in every
teacher's library."

Ed. Courant.—" It will help the teacher greatly.'

Va. Ed. Journal.—" The author draws from a large experience."